MOCKINGBIRD TRIO

Bird room in Queens

MOCKINGBIRD TRIO

ARLINE THOMAS

ILLUSTRATED WITH PHOTOGRAPHS

Charles Scribner's Sons • *New York*

TO CHUCK, ANDY, AND GEORGE—
MAY THEIR TRIBE INCREASE

 CONTENTS

ILLUSTRATIONS

MOCKINGBIRD TRIO

1

Three Baby Mockingbirds Arrive

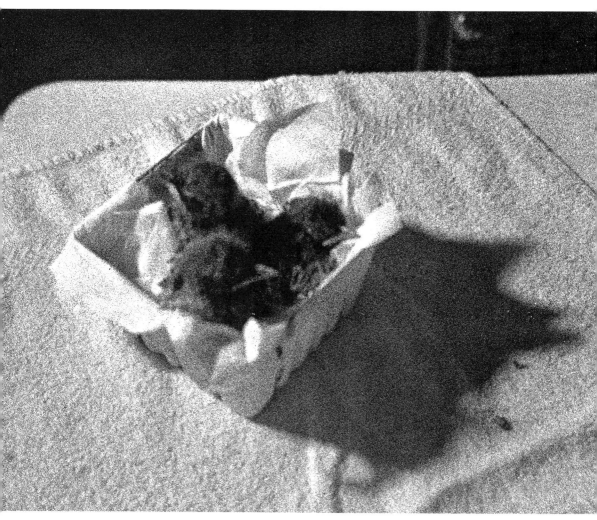

On arrival

1 THREE BABY MOCKINGBIRDS ARRIVE

"Chuck! Chuck!" Their food call exploded from the wide-open mouths of two baby mockingbirds. The third baby in the shoe box was quiet. The bearded young man named Andy looked worried. He had just rushed the birds in his car to my Long Island bird shelter.

"Why doesn't that one make any sound? Is it all right?" he asked.

I hurried from the refrigerator with the formula I keep on hand for orphan birds. "Let's get some food in them," I said. "It's been a long time since they've had any. The little fellow may be weak or too scared to open its mouth."

With a dash of hot water, I warmed and moistened the food. It was a mixture of grated hard-boiled egg, chopped beef, dry baby cereal, and powdered eggshell, plus wheat germ and vitamins. I feed it to the babies of insect-eating birds such as robins, mockingbirds, and thrushes. The young of many kinds of seed-eating birds (sparrows, finches, cardinals) thrive on it also. When they are in the nest, their parents feed them insects too.

Using blunt tweezers, I dropped a pea-sized morsel into the yellow mouth of one noisy youngster; the other gulped down the next tweezerful. My hand flew back and forth until I had fed each one four times. Then with stomachs full, they were quiet.

Now I could give my attention to the third one. As I picked it up, I saw it was smaller than its nestmates. It raised its head as if expecting food, but its beak stayed closed. Holding it in my left hand with its back against my palm, I curved my fingers around its breast. I felt its feet clutch my hand and it crouched there quietly. With my fingertips I opened its beak and slipped a bit of food down its throat. This is called force-feeding. As I did not want it to choke, I took care to give it much smaller pieces than I had fed the others.

Afterward I put the orphans in a small berry box lined with grass and tissues. This was the best substitute I had for the cuplike nest mockingbirds make from twigs and grass. The grass and tissues I used gave the feet of the little birds something to push against. If they had been put on a hard, flat surface their legs might have "spraddled." Because young birds grow so fast, their legs could set in that position. Then they might never be able to walk properly.

There were soft gray feathers on their heads and white ones with grayish spots at their throats. But there were still many spikes of rolled-up quills on their partly naked backs and wings.

To grow feathers baby birds need a great deal of nourishing food. Each feather starts in a small pit or hollow in the inner layer of skin. As many as 3000 or 4000 feathers are needed to cover a small bird. Feathers come in different sizes and shapes. They are nourished by the bird's blood, which is brought to each pit by a tiny artery. When a feather is full grown, the artery is sealed. The feather stops growing but it stays anchored in its socket for months. When a feather gets worn out by long use, the cells in the bird's underskin start making a new one. The old feather falls out and the new one takes its place. This is what happens when a bird molts.

Another reason young birds need plenty of nourishing food is that they have to manufacture strong bones and muscles in a hurry. A bird's body has many muscles. They are needed for flying. Sometimes the breast muscles, which provide power for the wings, account for almost half the bird's weight. In only two weeks after hatching, young garden birds like these mockers leave their nests and start walking and fluttering.

One thing baby birds do not need while they are in their nest is water. They get their moisture from their food. It is dangerous to give young birds water or milk from a medicine dropper. When they are old enough to perch, they drink water by themselves.

On their arrival at my shelter the three little mocking-birds not only were not fully feathered, they were even too young to perch. From this I judged that they were about ten days old. Probably it would have been three or four days before they would have left their own nest. While they were in it, their mother brooded them at night and often during the day. At breeding time her breast temperature is well over 110 degrees. Except for young nestlings, birds' temperatures are higher at all times than those of mammals. Birds also have nonconducting plumage which conserves heat and keeps their temperatures at a high level even in cold weather.

Baby birds do not have such a high temperature, nor do they have a full supply of feathers. So in caring for them, we have to keep them warm at all times. For this reason, I put a scrap of wool across their backs like a blanket. At night I set them, basket and all, in a small brooder. It has a plastic dome fitted with a light bulb. This sheds just enough heat to keep them snug and cozy. They need warmth especially during the night when they are not being fed. At that time they lose heat quickly.

While I was handling the birds, the always talkative Andy had been strangely silent: "Wow! my lunch hour's over, I'll have to run," he called out at last, heading for the door. As he left, his eyes twinkled. "Name the little one that doesn't open its mouth after me." Enjoying the joke, I did.

I had set my oven clock to ring in fifteen minutes. When it did I again force-fed the newly named Andy. Because the other two were such willing eaters, I fed them only every half-hour. Before the week was over, they had names too. Mildred Hearl, a friend who was taking pictures of the birds, suggested calling the one that seemed most alert and independent "George," after her capable husband. Chuck got his name because of his especially loud and lusty food call.

To give Andy a little extra pep, later that first day I baked a bird custard. This is made with one-third cup of water and a well-beaten egg but no salt or sugar. Although Andy still did not give their "chuck-chuck" food call, he swallowed the custard readily. But I had to open his bill with my fingers first. Chuck and George always opened their beaks for food. They were eager and noisy eaters. I could almost see them grow. Thanks to my constant force-feeding, Andy was growing too, but he had not caught up to them.

Babylike, they slept most of the time. When awake, they stretched and flapped their stumpy wings. They also spent a good deal of time preening. This helped pull off their quill casings. These are transparent sheathes which split and let the feathers open out. Bits of the casings fell on their backs. Sometimes they looked as if they had dandruff.

During the day, I took their berry box out of the brooder. I then set it in a cardboard carton that I had first lined with newspapers. This was done to keep the bottom clean. The efforts of even a tiny baby bird to avoid soiling its nest are amazing. Chuck, Andy, and George were no exceptions. First one or another would stir and suddenly raise himself on uncertain legs. Then he would turn round a full circle. Moving backward a few wobbly steps, he would lean forward and downward and stick his little tail over the edge of the berry box. After shooting his dropping over the edge, he would right himself. Then settling down contentedly under the blanket again, he would go back to sleep.

By their third day in the berry box, George decided the time had come to leave. Standing up, he perched on the rim. After a couple of wing flaps, he jumped bravely out. Landing in the middle of the carton, he looked all around. It's a big world, he seemed to be thinking, but, like his competent namesake, somehow he felt he could cope with it.

Playing follow the leader, Chuck jumped out right behind him. He too seemed to think it was a big world, but he was not so sure what to do about it. Running to a corner, he tried to hide.

Alone in the berry box, Andy peered over the edge at the others. Then he too perched on the rim, but with nothing to weight it down, the light box tilted. Losing his balance, he tumbled over and out, sprawling on his back. I had to pick him up and set him on his feet. His first venture as a fledgling into the world outside the nest was a flop. When young birds have enough feathers to leave the nest, they are called fledglings. Before then they are known as nestlings.

The little mockers were now promoted to a box with a perch. Clutching the stick with their dark feet, they sat on it most of the day. At feeding time, Chuck and George called loudly and opened their mouths wide. Tight-billed and mum, Andy waited. So, as I had done from the beginning, I picked him up and opened his beak with my fingers. After almost a week, the force-feeding operation was still going on.

What was the matter with him, I began to wonder. Had he grown to like the special attention I gave him at feeding time? Even newly hatched birds open their mouths at the touch of food in their mother's bill. This reflex action is linked to certain nerve cells. It is an inherited reaction and no learning is needed for it. When a baby bird is fed, there is a series of reflex actions. The opening of its mouth is followed by the reflex of gripping the food and then of swallowing it. Surely Andy's reflexes had worked while he was being fed by his parents. Otherwise he would not have survived.

At their next feeding, I sat him on a table and tapped his beak with my tweezers. "Look, Andy," I scolded, "get with it. I can't force-feed you the rest of your life."

Did he really understand me? It almost seemed so. Or maybe his dormant reflexes began to work again. For a moment he gazed steadily at the tweezers. Then with a series of "chuck-chuck-chucks," he toddled over to the edge of the table and finally opened his mouth to be fed.

2

Behind the Scenes
at the Bird Shelter

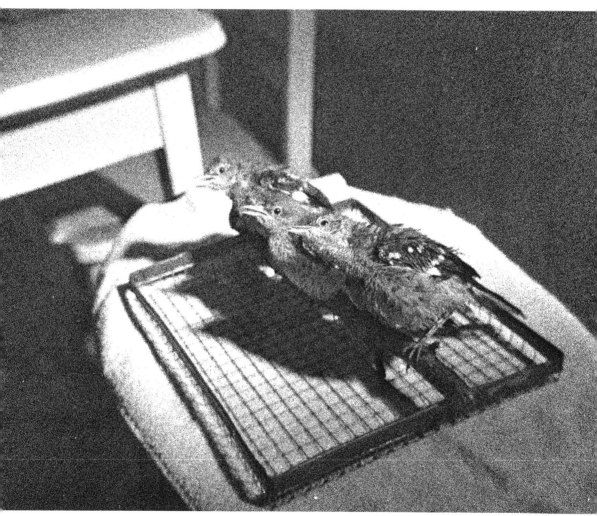

Sun bathing

2 BEHIND THE SCENES AT THE BIRD SHELTER

Anxious about his namesake, Andy phoned me the day after he brought the mockers. This young man is fond of birds and has rescued several injured ones. Sometimes he acts as my "bird ambulance" carrier. The ambulance is simply a cardboard carton tied with a string. It is used to bring helpless birds to my shelter.

The shelter is a little room eight feet square built onto my house like a porch. It is carpeted with newspapers and these are replaced every day to keep the floor clean. For furniture, it has a stick tied with ropes for the birds to swing on and two dead tree branches for them to perch in. Since the shelter is surrounded by high shrubs, the birds feel that they are really outdoors.

The room is stocked with food and water for the birds—often twenty or more of them—that I am keeping there. For the seed-eaters, there is a dish filled with wild-bird seed, peanuts, and lettuce. For the insect-eating and fruit-eating birds, I make a mixture of soaked dog-meal pellets, chopped beef, and hard-boiled egg. They get grapes, raisins, and cut-up bananas for dessert. Because the meat-and-egg mixture spoils quickly I put out only a small amount. I have to refill that dish several times a day. Sometimes when I am late getting home from a shopping trip, the birds empty the dish. When this happens, they greet me with a wild commotion as I walk up the steps. "What a nice welcome," my neighbor once said. "The birds seem so glad to see you."

Having heard their language before, I knew they were jeering, not cheering. "Yah! Yah! Yah!" yelled the jays. The starlings went into a tizzy, screaming "Fee-u, fee-u, fee-u" at me. "Tut-tut-tut," scolded the robins. Even the shy hermit thrush let out a shrill whistle. In bird talk this all added up to "Where've you been? Our dish is empty! Bring us food, food, food—and on the double."

In this aviary I keep injured birds that are on the mend. When they are well, I release them. Wild birds that have become tame may not survive when they are set free. After I put a bird in this room I never touch it again. By the time they are ready to go, the same birds that once ate out of my hand have become so wild that they have to be caught with a net.

People in my neighborhood often bring me birds because of my work with the Audubon Society. This society was named in honor of John James Audubon, who lived from 1785 to 1851 and spent his life studying and sketching the birds of the United States. His beautiful paintings, *The Birds of America*, show the many kinds there were in this country in his time.

The emblem of the society is the American egret, a bird it helped save from extinction. In the early part of the twentieth century the long white plumes of this bird, a variety of heron, were much used to trim women's hats. Plume hunters slaughtered thousands of egrets every year to get the feathers. Through the work of the Audubon Society and others, a law was passed to stop them.

Started in 1886, the society had only a few members at first. Now it has thousands in every state. Its aim is to save the country's wildlife and forests. It also tries to stop the pollution of water, soil, and air. Some of its members make studies of endangered birds, such as the bald eagle, the whooping crane, and the California condor, to learn about their histories and habits and what is causing their numbers to decrease. Others work with schools and colleges to show students ways to improve our environment.

Some members, like myself, take care of injured wild birds. This group is called the bird-saving corps. It was started because most animal shelters cannot give wild birds the special care they need.

At its headquarters in New York City, the society gets telephone calls from people who find injured birds. They want to help the birds but do not know how. When people in the suburbs call animal shelters about birds, they are told to call Audubon headquarters. Then the society gives them the name of an Audubon volunteer in their neighborhood.

Anyone doing this work must have a bird-salvage license from the state and the federal government. In New York State it is against the law to keep wild birds without such a license. The Audubon Society members who do such work must know about first-aid for injured birds. They also must be able to tell what kind of bird they are caring for. Then they will know what it eats in the wild and what substitute food to give it. My freezer is stocked with raw beef for hawks and owls, fish for gulls and terns, and hamburger for insect-eating birds.

Wild-bird care is a combination of common sense and know-how. A beginner can find little help in books; few have been written about the subject. Perhaps this is because in some states there are laws against keeping wild birds without a permit.

In big cities birds get into many kinds of trouble. During migration, some birds crash into skyscrapers. Others get exhausted in stormy weather and fall to the ground. Unless rescued and given proper care, they are almost sure to die.

In the suburbs, birds also get into trouble. Starlings build nests in air conditioners and have to be removed. Sparrows and finches set up housekeeping in rolled-up awnings. When the awning is lowered, their babies fall to the ground. Birds flounder into swimming pools, get hit by cars, and fly against overhead wires. When caught by cats, they may die of infection unless they get antibiotics.

In spring and summer baby birds fall out of their nests. This is when many telephone calls pour in from frantic people. "What should I do?" they ask.

If very young baby birds fall out of their nests, the best thing is to put them back. If they are old enough to hop or flutter, it is a good idea to perch them in a tree or bush out of harm's way. The parents are usually nearby. They do a better job of raising their offspring than people do. Being a foster parent for young birds takes a great deal of time and patience.

Every summer I get dozens of young birds. Some of them are charmers, like one three-day-old mallard duck. Unfortunately it had been too long without food and did not survive. But that same year a nestful of robins, two Baltimore orioles, and an assortment of jays, starlings, and sparrows all lived to grow up. When they were old enough to take care of themselves I set them free.

"Do the young birds you raise ever come back?" people often ask me. I make sure they do not. Taking them to a distant woods, I release them where there is plenty of food and cover. In this park berried shrubs and trees have been planted to supply these. The birds are safer there than in residential neighborhoods where dogs and cats often harm birds.

The baby mockingbirds were found by a park-department man. Someone had put them in a shoe box and left them near the park's entrance.

It was early in August when I got the message about the three little mockers. At that time I was nursing a number of other birds at home. Unable to go for the orphans, I sent an SOS to Andy. He obligingly dashed over to the park in his car and brought them to me.

During the summer when I am caring for young birds, my friends, George and Mildred Hearl, often come over to take pictures of them. The day they came to photograph Andy, Chuck, and George, they brought a floodlight. They decided to use it instead of flash bulbs since those do not always work.

Putting the birds on a stick in the kitchen, they turned the brilliant light on them. Suddenly all three birds fell over sideways like a row of dominos. They fluffed their feathers, stretched out their legs, and held their wings at crazy angles.

For a moment we thought they were having a fit. Then we realized they were doing a sunning display. Birds often do this when they are touched by a shaft of sunlight. They seem to have no control over their actions. They stop suddenly, fan their tails, and lie with outspread wings on the ground or on a branch. While they are doing this, they keep one eye cocked toward the sun.

One explanation for this odd behavior is that it is the

birds' way of getting vitamin D. The preen oil that coats a bird's feathers contains ergosterol, a pro-vitamin compound which changes to vitamin D when exposed to irradiation by sunlight. When it preens itself after a sunning display, the bird swallows tiny globules of this oil. In this way it gets vitamin D from its sun-warmed plumage.

As soon as the floodlight was snapped off, Chuck, Andy, and George straightened up. When it was turned on again, they all keeled over as before. Sun or electric light was the same to them. All they wanted was to bask in the warm rays.

3

Seeing Snakes

Safe at home

3 SEEING SNAKES

During the time that Chuck, Andy, and George were being hand fed, I kept them in the kitchen. Their carton, covered with wire netting, rested on a low stand near the refrigerator. In this way, they were always near their food base.

One morning I had to press some clothes in the kitchen. Pulling down the ironing board, which was close to their box, I plugged in the electric iron. As it slid back and forth on the board, its black cord coiled and uncoiled over the birds' heads.

To my utter surprise, they began to panic. Chuck-chucking madly, they raced around and around in their box. One minute they seemed to be trying to hide in a corner. The next they were clawing at the wire cover, trying to escape.

What had got into them, I wondered. Was a bee or a wasp in their box? But even little birds are not usually afraid of these insects. Rather it is the other way round. Bees and wasps make nice snacks if they come within pecking distance.

I stopped ironing to search their box. Instantly the commotion stopped. Just some little game they were playing, I thought, one bird chasing another.

Again I started to iron. This time the birds grew so frantic that I was alarmed. They threw themselves against the top of their box so hard I was afraid they would bash their heads. As though panting with fright, their beaks kept opening. I picked Andy up to try to soothe him. His heart was pounding and his eyes were glassy with terror.

Whatever it was, something in the kitchen had frightened them. Puzzled, I carried their box into another room. They quickly calmed down and began their interrupted preening.

Back at my ironing board, I thought about their odd behavior. That electric cord, coiling and uncoiling over their heads, had they mistaken it for a snake? But these little fellows were only ten days old when I got them. They could never have seen a snake. Snakes in this part of Long Island are as rare as snowstorms in July. This then, I reasoned, must be an inborn fear, the kind the animal behavior experts write about. It is one which they believe is handed down from the ancestors of birds and animals for their protection.

Later I studied a copy of one of the famous paintings in Audubon's *Birds of America* series, plate 21. It shows four mockingbirds fighting off their mortal enemy, the rattlesnake. In a yellow jasmine bush, the rattler is coiled about a nest. Jaws gaping, it seems about to devour the eggs of a nesting pair. One of the birds has lit on the snake's back and is trying to peck one of its eyes. The other is perched close to the murderous mouth. The snake's forked tongue is lashing at the defender of the nest. Two other mockers, perhaps neighbors, hover above the snake. One seems to be eying the rattle, which even in the picture appears to be quivering. There can be no mistaking the frenzy of those painted birds. Mockers and snakes are natural enemies.

Roy Bedichek, a Texas naturalist, has told about a battle

32

he saw between a mockingbird and a rattler a few years ago. The mockingbird darted from above and flew along the back of the snake until it reached its head. Angered, the rattler coiled and struck. Then it began to rattle. By this time one of its eyes was bleeding. Perched above it on a limb, the mockingbird waited. When the snake began to glide away, the bird went into action. Showing no fear for its own life, the mocker swooped again and again at its enemy. At this point the snake seemed frightened. It tried desperately to escape. The mockingbird gave it no chance. Each time the bird flew back to the limb, it did a sort of war dance. Slowly opening and closing its wings, the bird hopped to one side and then the other with mincing steps. Then it dive-bombed its age-old foe with fury. Finally it struck the snake's other eye. Blood spurted from it as the bird kept jabbing at the socket. Frantic and now totally blind, the rattler coiled and struck in all directions. At last, maddened with pain, the sightless reptile sank its fangs into its own body. In the end it died from its wounds.

Snakes eat many rodents, but they also like birds' eggs and young birds. They get these by crawling up the tree in which a nest is built. Most garden birds, although they scream and scold, are helpless against snakes. Not so the plucky mocker; it gives battle.

Naturalists report that a pair of mockingbirds will often attack a snake that tries to rob their nest. They seem to have a prearranged plan of war. One of them will hover over the snake and strike at the back of its head or neck. The other flies in the snake's face with fluttering wings to draw its attention.

The attacking bird seems to know enough to strike always at the same spot. Dead snakes have been found under mockingbird nests with only a round hole in the back of the head.

It would seem that once the bird draws blood, it keeps on striking at the reddened spot. Scientists are still puzzled at the mockingbird's actions with snakes. Are they caused by transmitted knowledge or protective instinct?

Chuck, Andy, and George, forgive me. Your fears were real, although the snake was not. Instead of chasing it away as a parent mockingbird would have, I let it come near you without lifting a finger. As a bird foster parent, I have a lot to learn.

4

Family Tree

Bird bath

4 FAMILY TREE

Visitors to my home that summer were often surprised to see Chuck, Andy, and George. "Mockingbirds in New York?" they questioned. "Aren't they a southern bird?"

Mockers were and still are a symbol of the Old South. In fact, they are so well liked in the region that they have been named the state bird of Mississippi, Florida, Texas, Tennessee, and Arkansas. There are many of them in all these states, especially Florida.

Mockingbirds, however, have a pioneering spirit, always pushing out and extending their range. They will go wherever they can find trees and shrubs 10 or 15 feet high. A good food supply is also very important.

In the last twenty years, some have been coming north and staying the year round. They are now found in the New England states, New York, and New Jersey. Scientists are not sure why they have come but they are most welcome. Everybody loves a mockingbird, it seems. If a poll were taken of the favorite bird in this country, the result would be a toss-up between the mockingbird and the robin.

More than 200 years ago, a naturalist named Mark Catesby discovered the mockingbird. He saw it for the first time in Charleston, South Carolina, and made note of it. Afterward he sent a report to Carl Linnaeus, a famous Swedish scientist. At that time Linnaeus was listing and naming all the known plants, animals, and birds in the world. Delighted to learn about the mockingbird, he added it to his list of birds.

What kind of a bird is this all-time favorite and how does it live in the wild? Belonging to an even more thoroughly American family than the wrens, mockers are closely related to them. There are about thirty species of mockingbirds and their relatives, the catbirds and thrashers. Most of these birds are fine singers, and many are good mimics, the mockingbird especially. The name of the family is Mimidae, and mockingbirds are sometimes called "mimine thrushes." They probably developed, as did the wrens, from some thrushlike ancestor. The scientific name of the mockingbird is *Mimus polyglottos*, meaning "many-tongued mimic." An Indian name for it is "bird of four hundred tongues."

Early in spring mockingbirds start their courtship. It begins with a sort of square dance. The male and female birds face each other, wings drooping, heads and tails high. They swing their heads as if they were going to hit each other with their beaks. Afterward they step together sidewise, only going a few inches at a time. When they have moved five or six feet, they wheel and go back the same way. After each dance, the birds fly off in opposite directions for a short distance.

Soon after this courtship dance, the pair stake out a territory. The male will never let another male mockingbird near it, but he does not always chase other kinds of birds. When he does, he rivals the kingbird as a fighter. At times it seems as if he goes around looking for a fight. This is by no means limited to nesting time. Mockers fight even among themselves, and six or eight pairs sometimes have a battle royal.

Most birds sing from a perch, but the mocker can sing while flying. Sometimes it interrupts its song to drive off another bird. It cannot stand the sight of a robin. Some authorities think that the rusty red shade of the robin's breast is what upsets it. Another reason may be that robins and mockers eat more or less the same kind of food. Perhaps the mocker thinks that the robin will cut down its own food supply.

The naturalist A. C. Bent reports that a mocker which lived in his yard for days fought its own image in a cellar window. It probably mistook the reflection for an intruder muscling in on its territory.

Mockingbirds have a good homelife. The mates share burdens and dangers. Working very fast, they build their nest together. Each stick seems to be chosen to fit so that little time is needed to put it in place. The sticks are laid down in layers. One end is pushed below two or three other sticks. The other end is raised over one and under another. The building is almost like basket weaving. When the outer walls are in place, the lining of lighter material is worked in. This is often Spanish moss in the South, or feathers, horsehair, thistledown, or tufts of cotton. The materials are pressed down and smoothed evenly. The inner lining of the cup is always of softer materials. It is made to cradle the eggs without piercing them and to protect the nestlings from injury and from the elements.

The composition of the nest may be partly from memory. The parent bird makes a nest resembling in place, materials, and size, the one in which it lived as a nestling. But where, except from inherited instinct, does the actual knowledge of construction come from? If they breed in the first year after hatching, as many mockingbirds do, they have never watched a nest being built. They have never taken lessons in the way to buttress and anchor a foundation. They have never been taught how to gather and interweave the materials. No one tells them how to fashion and line the inside. This know-how, like so many behavior patterns of birds, must surely be an instinct handed down from parent to offspring. They build their nests, sing, dance, and bathe as an expression of inheritance. Chuck, Andy, and George certainly loved to bathe. They dunked and splashed around in their water dish until it was empty.

After the spotted blue-green eggs are hatched, the father gets nervous and bad-tempered. He fusses at everyone who comes near the nest. Sometimes he will scold at a person coming along the path.

The father mockingbird is a model family man. He protects his home from intruders and helps feed the nestlings. During this time he is so busy that he does not sing much in the daytime. He makes up for this on moonlit nights when he pours out his carols and trills. Sometimes the female will answer with a "twit" or "tweet." While on the nest she does not sing at all.

No policeman ever walked a beat better than the father mocker does. He pays particular attention to leaning trees or places where a cat could approach. But no cat can get at the nest of this policeman if he sees it. Like a flash the bird will strike from above with his sharp beak. Then he makes a scolding sound and the fight is on. His mate will make the same sound and join him in battle. They strike at the cat from different directions, always staying out of reach of its claws. Even a large cat, when it is in a tree, is no match for a pair of mockers, as the cat soon realizes. When it jumps to the ground, they will drive it to shelter.

Ecologists, who study the relationship between creatures

and the environment, point out that mockingbirds are very use-
ful. They eat tons of destructive insects in spring and summer and
also feed these to their young. They seem especially fond of the
boll weevil, a beetle that destroys cotton crops by attacking the
cotton boll or pod. Cinch bugs, caterpillars, and crickets are also
part of their diet. They are also fond of berries.

As youngsters, Chuck, Andy, and George had smoky
gray backs, and below their spotted chests were white vests.
Mature mockingbirds seem almost all gray except for two white
wing bars. In flight these bars unite to form a broad white patch.
The outspread tail shows white feathers in a V-shape toward the
rump.

Mockers have a habit of raising their wings and holding
them high before folding them. They repeat this opening and
closing of the wings, fanning them gently. When two or three
mockers are seen doing this at the same time, it looks like a bird
gymnastic drill.

Chuck, Andy, and George sometimes put on a little show
of their own. As is characteristic of mockers, they ran around the
floor, opening their wings a little at a time by sudden jerks. At
such times they looked like three little mechanical birds that had
been wound up with a key.

5

Listen to the Mockingbird

Watching

5 LISTEN TO THE MOCKINGBIRD

Listen, dearest, listen to it,
Sweeter sounds were never heard!
'Tis the song of that wild poet,
Mime and minstrel—Mockingbird.
ALEXANDER BEAUFORT MEEK

The mockingbird is sometimes called "the poor man's symphony." Americans appreciate its concerts more than the song of any other bird. The naturalist R. A. Selle called the mocker "El Troubador." He says its song soothes insomnia. Describing it as a restless spirit expressing itself, he states that its voice trembles with feeling and yearning. Edward Howe Forbush, another bird expert, says that the mockingbird is the American king of song. It equals and even excels all others. Improving on most of the notes it imitates, it revels in the glory of its voice. Those who believe the mocker only mimics other birds' songs have much to learn of its wonderful originality.

The mockingbird's own song is heard at its best during the mating season. Then the bird flutters into the air from a tree-top and makes up its melodies as it goes along. It pours out all the power and energy of its being in such an ecstasy of song that it seems to exhaust its strength. Afterward it sometimes flutters down through the branches until it touches the ground. On moon-lit nights at this season the mocker fills the air with exquisite swells and trills. Watch a mockingbird some spring morning. With ruffled feathers and drooping wings, it sits on a bough and pours out its song. At times the intensity of its notes seems to lift it high in the air. With dangling legs and flopping wings, it drops again to its perch, singing all the while. Sometimes it drops to the ground for a moment. Still singing, it makes a few quick hops in the grass. Then it bounds into the air again, with scarcely a break in its song.

Music high and low, loud and soft, pours from its throat. With never a hesitation, never a false note, this gray-and-white troubadour serenades the world and its mate. At times its singing may be heard until dawn.

What gives the mocker its wonderful voice? Like that of other songbirds, its song starts in the syrinx. This is the seat of song in birds. It is a boxlike structure at the lower end of the windpipe. It contains two sets of membranes controlled by nearby muscles. Air currents from the lungs stretch and relax the second, more delicate set of membranes. This provides the variety of notes that some birds are able to sing. Mockingbirds seem to be gifted with extraordinary voice muscles.

Mockingbirds sing all year long, although much less in winter. They usually pick an exposed place to sing from—a roof, eave, telephone wire, or the top of a fence post. During mating time, mockers sing their own burbling melodies but many of their songs are imitations of other birds they have heard. They usually repeat each phrase three or four times before changing tune.

There seems to be no songbird that the mocker cannot imitate. One amazing bird, known as the Arnold Arboretum Mocker, lived in the Arnold Arboretum at Boston. Employees at that famous botanical garden stated that it could imitate thirty-nine birdsongs, fifty bird calls, and the sounds of a frog and a cricket.

Mockingbirds are quick to pick up any new song they hear. Some have given perfect imitations of thirty different kinds of birds in succession. They also imitate correctly such unbirdlike sounds as the postman's whistle, a squeaking wheel, and the buzzing of locusts. They can imitate crowing roosters, cackling hens, and barking dogs. At times they even mimic their own offspring.

In America the mockingbird has no singing rival, but in Europe and Asia a little six-and-a-half-inch thrush has been called by poets the most famous of all songsters. This is the nightingale, a modest-looking reddish-brown bird that is very shy. For centuries the nightingale's song has been celebrated in verse and story.

Some years ago several nightingales were brought from Europe in cages and housed in the gardens of the Bok Singing Tower in Florida. The local mockingbirds flew around, inspecting the newcomers. For a while they listened to the magnificent song of the little brown bird. Then they began giving their own nightingale concerts. They imitated to perfection the lovely notes of the foreign bird. Records of the imitations were made with a sound spectrograph. (This is an instrument that makes a representation of sounds.) They proved to be exact reproductions of the original nightingale songs. They were perfect renderings of each phrase in all its parts, including vibrations beyond the range of the human ear.

Certainly the name "many-tongued mimic" is a good one. A bird lover once tried a poor imitation of a whippoorwill's song. A nearby mocker first repeated the imitation, then quickly followed it with a perfect whippoorwill call.

In Washington, D.C., a mockingbird sang along with the National Symphony Orchestra during outdoor concerts. The bird's big moment came when the flutist performed in "Peter and the Wolf." The mocker mimicked the flute that plays an imitation of the bird calls in that well-known fairy tale.

A Texas mockingbird is reported to have found a way to tease an electronics expert. This man had rigged his sliding garage doors to open and close when he whistled a certain tune. The bird would fly to the garage, perch on a convenient basketball hoop, and whistle the door up and down, up and down, up and down. . . .

A radio amateur in Dallas was practicing the Morse code. Later he began to hear messages outside his window at odd times. The sender turned out to be a mockingbird beeping out the "dit-dit-dahs" in perfect Morse rhythm.

My fledgling mockers were still too young to make any sound except their call note "chuck-chuck." However, I wondered if, later on, when they were set free, the weeks spent in my home would have an effect on them. In the middle of a lovely trill, would Chuck stop short to mimic a whirring eggbeater? Would George chime like my front door bell? Would Andy, perched in some high tree, suddenly startle his listeners with his imitation of a jangling telephone?

6

The Trio on Television

Feeding time

6 THE TRIO ON TELEVISION

Chuck, Andy, and George had been with me just two days when they made their first public appearance. It was at a talk that I gave at the library in Port Washington, Long Island. Since some of my listeners were children, I thought they would like to see what baby mockingbirds look like.

Bundling the birds into a clean berry box, I tucked their blue wool blanket around their backs. During the trip, the box was kept in a covered carton to avoid drafts.

The audience listened respectfully while I talked and showed colored slides of birds. When I brought out the little mockers, the children came alive. They crowded around the platform while I fed the birds with my tweezers. Since it was a warm day, I took the blanket off to let the youngsters get a good look at the little mockers.

One small girl stroked the always hungry Chuck. Thinking it was feeding time again, he opened his beak wide. "He's got a big mouth—sort of like a dragon," she whispered.

"Can I hold one?" another youngster asked. I took George out and sat him on the boy's outstretched hand, taking care not to let go of the bird myself. "This is swell, I like all kinds of wildlife," the twelve-year-old naturalist confided.

After a few more strokes and pats from the children, the birds were packed away and taken home. Their first public appearance had been a rousing success.

Toward the end of August I was invited to give a talk about birds on an afternoon television show. This was the popular Mike Douglas program, filmed in Philadelphia but with a nationwide hook-up. Since a live animal or bird always adds interest, the producer asked me to bring one of my birds.

By this time Chuck and George no longer ate from my hand. They were picking up their food by themselves. But Andy was still being fed with the tweezers. Since I wanted to show how a foster parent handles young birds, I decided to take him and feed him on the program.

Andy and I made the trip to Philadelphia by train. He traveled in a covered cardboard box with holes punched for ventilation. The box also held a jar of his food and my tweezers. When the train started, Andy gave a series of loud chuck-chucks. Maybe he was protesting his imprisonment or the unfamiliar motion. After all, it was his first train ride.

Then he grew quiet, almost too quiet, I thought. When we had traveled more than an hour and he had still made no sound, I began to worry. Had something happened to him? Was he getting enough air? Although I knew he might fly out, I decided to risk opening his box. Slipping off the cord, I carefully lifted one end of the flap and peeked in. Curled up against his food dish, Andy had his head tucked under his wing. He made the entire trip to Philadelphia fast asleep.

At the studio, I was briefed on what to do and given a small table to hold the food dish and tweezers. Mike Douglas discussed bird care with me for a while and then Andy was brought on.

Taking him in one hand, I picked up a tweezerful of food with the other and held it out to him. After his trip I was sure he would be hungry and eager to eat. Keeping his bill shut tight, Andy lifted his head and gazed at the studio audience. I tapped him on the beak, hoping to interest him in eating. Ignoring the food again, he switched his attention to the piano player. Several million people watch this program, but that made no difference to Andy. "Open your mouth, Andy," I pleaded. It was like old times in my kitchen when he first refused to eat unless I forced him. His reflexes must have gone haywire again, I reasoned.

A bit flustered by then, I put down my tweezers and picked up some food with my fingers. Quickly I pried Andy's beak open and pushed it down his throat. "This is how you feed a young bird when it won't open its mouth," I said. Luckily Andy swallowed the food. In a jiffy he was whisked off the screen. On his first and only television appearance, Andy bungled.

A week later I was asked to appear on a local television show. As before, the producer suggested I bring one of my birds. Because George was so poised and sleek, he was chosen to go this time. During the show, the host, John Bartholomew Tucker, mentioned that he had been surprised to learn we had mockingbirds in our area. He said he was not sure he would recognize one. At those words, George wheeled and flared his white wing patches. "Take a good look at me," he seemed to be saying. "From now on you'll recognize a mocker when you see one."

Chuck was the last to go on the air. He made his appearance on Patchwork Family, an educational program for children. Long-legged and long-tailed like all of his kin, Chuck was a handsome bird. He looked quite grown-up now.

During the show, as I discussed his family's traits, he gazed directly at the camera. In a close-up, there was no mistaking the eye stripe that gives the mocker its stern and penetrating look. At the end, I said good-by to the audience and added, "Chuck says good-by too." As though on cue, Chuck slowly dipped his wings. While his image faded from the screen, it looked as though he was waving good-by.

One by one, the trio had had their turn on television. Without any coaching, Chuck and George seemed to know how to put on a good show. Not so Andy. He would never get an award for the way he performed.

7
Good-by

Banded for departure

7 GOOD-BY

By mid-October the hickory trees around my house had already changed color. Their leaves, golden now, hung over the bird shelter so that even on cloudy days the sun seemed to be shining in there. Solitary robins and thrushes pulled the red berries from the yews in my yard. They were fattening up for their long trip south.

Chuck, Andy, and George were now more than two months old. Their speckled breasts were changing to pale gray. All the downy feathers of babyhood had gone; their bodies were now well defined and compact. They looked so much alike it was hard to tell them apart. Only their personalities were still different.

Chuck and George had always been less dependent than Andy. Now they were aloof and almost wild. When I walked into the bird shelter, they flew to the other end, out of arm's reach. Never, never, would they let me touch them.

For quite a while Andy had been picking up his own food as the others did. Sometimes, however, he took a tidbit from my fingers—a throwback to his younger days of handfeeding.

Now I stopped giving him anything from my hand. I wanted him to become as wild as the others. He had to be wary of everyone, including me. When he was released, this would be his protection.

Andy's playfulness also made him seem younger than Chuck and George. He teased the three bluejays that shared the shelter by running up behind them and tweaking their tails. When they turned and screamed at him, he seemed to enjoy the commotion he caused. He teased Chuck by pretending he was going to snatch a grape Chuck was eating. When Chuck flew off scolding, Andy acted as though it was a game.

As the days got shorter, all three mockers sat for hours by the windows in the shelter, gazing into the outside world. Certainly they were no longer the tame little birds I had fed in my kitchen. A change had come over them. They acted as though they would like to be on the move. I knew the time had come to let them go. Once freed, they could either go south or spend the winter on Long Island.

Before I released them, however, I wanted to have them banded. This is a practice that dates back to the Roman Empire, about two thousand years ago. The marking of birds was started to identify the falcons of the emperors. Modern bird banding began in Denmark in 1890. A school teacher named Hans Christian Mortensen put metal bands on the legs of some ducks, storks, and hawks. The bands had his name and address on them. Later the banded birds were found in many places on the continent. Then other bird students started "ringing," as it is called in Europe.

In a short time, bird banding caught on in America. At first it was done by a group called the American Bird Banding Association. In 1920 the United States Fish and Wildlife Service took over the work. Since then banding of migrating birds in the

United States and Canada has been under this service's direction.

Birds are banded to give information about the individual bird and also about the species. Thanks to bird banding, it is now known that many wild birds live as long as ten years, some even longer. A red-winged blackbird banded in New York was found fourteen years later in North Carolina. Some Canada geese live to be more than twenty years old. The longest a North American banded bird has been known to live in the wild is thirty-six years. The bird was a herring gull banded in Maine in 1930 while still in the nest. In 1966 it was found dead on the shore of Lake Michigan.

From banding, scientists have learned that the golden plover returns north over a different route from the one it follows south to its wintering grounds. Banding has shown that the Arctic tern makes the longest migration flight. This was learned from bands returned from France and from Natal and Cape Province, South Africa. This bird sometimes makes a yearly round-trip flight of about 25,000 miles. It nests north to the Arctic Circle, and some reach the oceans near Antarctica in winter.

Anyone over eighteen years old who can identify all the common birds may apply for a banding permit. However, the applicant must be vouched for by three other bird banders. Persons doing this work get bands free from the Bird Banding Laboratory. They also get the forms needed for keeping their records. When a band is put on a bird's leg, the bander records the number and the kind of bird. He also notes the bird's age and sex, and the place and date of banding. Later he sends his record to the Banding Laboratory. There this information is put in a computer where it can be released when the band is sent in. Over a million birds are banded every year. Of this number, more than a hundred thousand bands are recovered.

Special traps or nets are used to catch the birds for banding. The bander must be very careful not to injure the birds. He must visit his traps every hour so that the birds do not suffer from exposure. When the trapped birds are removed, they are identified and examined for age and sex. Then they are carefully fitted with aluminum bands and released.

There are seventeen different sizes of bands. Large bands are used on swans, geese, and eagles. There are very small ones for tiny birds such as warblers and kinglets, and even one, the smallest of all, for hummingbirds. Besides its number, each band reads: "ADVISE BIRD BAND, WRITE WASHINGTON, D.C."

During one of my talks about bird banding, a youngster looked puzzled. "Do you cut the bird's leg off that's banded and send it to the government?" he asked. I assured him that this was not necessary and explained what should be done. Anyone finding a banded bird should first take the band off, straighten it out, and tape it to a piece of heavy paper. With the band, give the finder's name and address and the date found. Be sure to give all the numbers and letters on the band. The finder should send all this information to the Bird Banding Laboratory of the Fish and Wildlife Service. The address is Laurel, Maryland 20810. If a live banded bird is found, the band should not be removed. The finder merely has to write down the number on the band and then release the bird. This information should also be sent to the Bird Banding Laboratory.

Near the end of October, Chuck, Andy, and George were completely wild and ready to leave. A bird-banding acquaintance came and tagged them for me. The next day I drove with them to the nearby Jamaica Bay Wildlife Refuge. Here there are various berried plants, including autumn olive and wild rose bushes.

These are grown to furnish food in the fall and winter for many kinds of birds, including mockingbirds.

When we arrived, I walked to a secluded spot near a grove of pines. There I opened the carton I had brought them in. To my surprise, Andy was the first to leave. He flew to a nearby branch and looked all around him. Seconds later Chuck and George streaked off to a clump of shrubbery. Andy hurried to join them. Wishing them happy landings, I followed their flight by the flicking white of their tail feathers. That was the last I saw of the trio.

May it be a long, long time before bands 762-56434, 762-56435, and 762-56436 are sent in.

Andy

INDEX

Index